A PEEK AT ROOTS

DOUG BRADLEY

PowerKiDS press

Published in 2026 by The Rosen Publishing Group, Inc.
2544 Clinton Street, Buffalo, NY 14224

First Edition

Editor: Greg Roza
Book Design: Leslie Taylor

Photo Credits: Cover, p. 1 AngieYeoh/Shutterstock.com; p. 5 brgfx/Shutterstock.com; p. 7 VVproduct/Shuttertock.com; p. 9 Filipe B. Varela/Shutterstock.com; p. 11 ER_09/Shutterstock.com; p. 13 imageBROKER/Thomas Filke/Alamy.com; p. 15 Tayfun Sertan Yaman/Alamy.com; p. 17 Rafael Novais/Shutterstock.com; p. 19 (turnips) Lahore Qalandars7/Shutterstock.com, (carrots) alicja neumiler/Shutterstock.com, (radishes) Uryupina Nadezhda/Shutterstock.com; p. 21 Tint Media/Shutterstock.com.

Library of Congress Cataloging-in-Publication Data

Names: Bradley, Doug, 1971- author.
Title: A peek at roots / Doug Bradley.
Description: New York : PowerKids Press, [2026] | Series: A peek at plants
  | Includes index.
Identifiers: LCCN 2024053782 (print) | LCCN 2024053783 (ebook) | ISBN
  9781499452044 (library binding) | ISBN 9781499452037 (paperback) | ISBN
  9781499452051 (ebook)
Subjects: LCSH: Roots (Botany)–Juvenile literature.
Classification: LCC QK644 .B646 2026 (print) | LCC QK644 (ebook) | DDC
  581.4/98–dc23/eng/20250107
LC record available at https://lccn.loc.gov/2024053782
LC ebook record available at https://lccn.loc.gov/2024053783

Manufactured in the United States of America

CPSIA Compliance Information: Batch #CSPK26. For Further Information contact Rosen Publishing at 1-800-237-9932.

Find us on 

CONTENTS

The Parts of Plants

Our world is filled with plants of different sizes, shapes, and colors. Most plants have the same parts. Stems help plants stand tall. Leaves make food for plants. Flowers and fruits look pretty, and they allow seeds to grow. What about the roots? Read on to learn all about them!

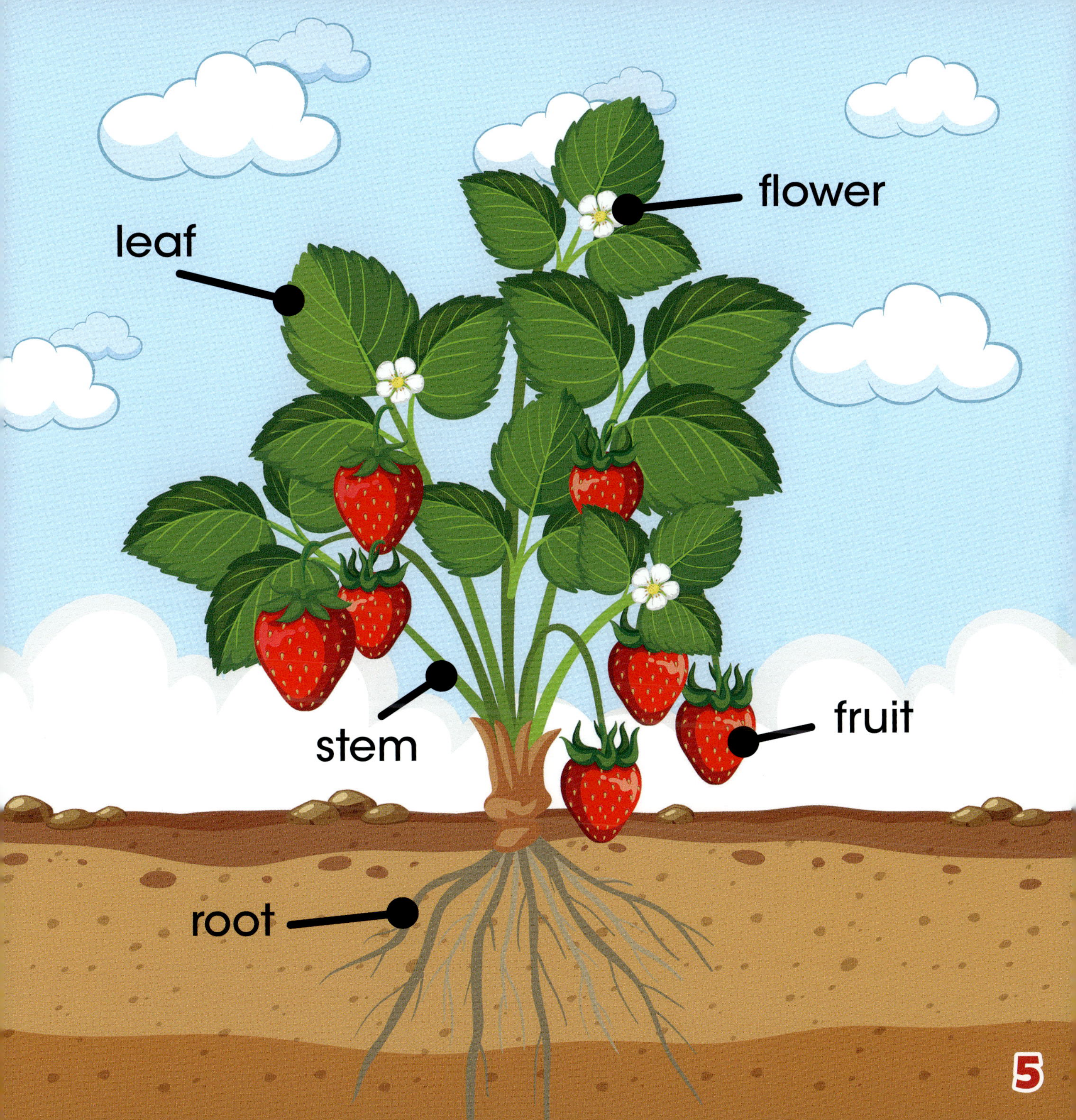

leaf
flower
stem
fruit
root
5

Sprouting Roots

Most plants start as a tiny seed. A seed starts to grow when it's in the ground. The first thing that happens is that a **sprout** begins to grow downward. This sprout forms roots as it grows. These roots **branch** out and keep the plant in the ground.

Water Break!

Roots help plants stay in the ground. But they have another important job. Roots take in water from the ground. They also take in **nutrients** the plant needs to live. Water and nutrients move through the roots to the stem and then to the leaves.

Root Systems

There are two kinds of root **systems**. **Fibrous** roots are made up of many branches reaching into the ground. These roots are usually very fine. A plant's fibrous roots are mostly all about the same size and length. Grass has a fibrous root system.

11

The other type of root system is called a taproot. A taproot is one thick root that grows down. Other smaller roots may branch out of the taproot. Taproots grow deep and keep the plant firmly in the ground. Many trees have taproots. Carrots *are* taproots!

13

Mangrove Roots

Not all roots are hidden underground. Mangrove trees grow on coasts in warm places. They grow in salty water. They have large roots you can see. These roots help keep the trees in the ground even when waves hit the trees. Mangrove trees keep coasts from washing away.

Pando!

Some plants share a root system. The largest root system in the world is in Utah. About 47,000 aspen trees all grow from the same root system! **Scientists** think these aspens, known as Pando, are one big tree. *Pando* is Latin for "I spread."

17

Root Farming

Farmers know how to care for plants and their roots as they grow. Farmers put water and nutrients in the soil for roots to take in. Some farms grow root vegetables, such as carrots and beets. Other root vegetables include radishes and turnips.

Start Your Own Garden!

You can grow your own root vegetables! You can grow them in a garden or in a large pot. Root vegetables need a lot of dirt to grow down into. If not, they could turn out too small. What root vegetable would you grow in your garden?

GLOSSARY

branch: To grow off of a larger plant part. Also, a smaller part of a plant's stem that grows off the stem and often grows leaves, fruits, and flowers.

fibrous: Like thread or string, long and thin.

nutrient: Something taken in by a plant or animal that helps it grow and stay healthy.

scientist: Someone whose job is to explore things to discover more about the world.

sprout: A new growth from a seed or plant.

system: A group of objects or parts that work together.

FOR MORE INFORMATION

BOOKS

Connors, Kathleen. *How Do Carrots Grow?* Buffalo, NY: Gareth Stevens, 2022.

Kirkman, Marissa. *Roots.* North Mankato, MN: Pebble, 2020.

Silvora Collections. *Plant Parts and their Purpose: A Budding Botanist's Activity Book.* Independently published, 2024.

WEBSITES

Structure and Function of Roots
www.youtube.com/watch?v=K0_tAHBdXec&t=2s
This short, charming video teaches all you need to know about roots!

INDEX